CELLULAR RESP

Death and Destruction with a side of alpha-ketoglutarate

By William Brucker, Kate Schapira, Brenna Brucker,
Christoph Schorl, Ava Lovato, Stacy Croteau, Carolina Veltri
Illustrated by Grayson Armstrong, Cybele Collins, Andrew Peterson,
Chris Boakye, Joao Paulo, Ivy Bradley, Aziz Khoury
Edited by Melina Packer, Kate Schapira, Brenna Brucker

For information about permission to reproduce sections of this book, email:
providencealliance@gmail.com

Library of Congress Cataloging-in-Publication Data

Cellular Respiration: Death and Destruction with a side of alpha-ketoglutarate
Library of Congress Control Number: 2011963045
Brucker, William
ISBN: 0982818904

Printed in the United States

Acknowledgments

Peterson Foundation, Donald and Sylvia Robinson Family Foundation, Dr. Daithi Heffernan, Dr. Andrew H Stephen, John Luo, Melina Packer

Index of Concepts

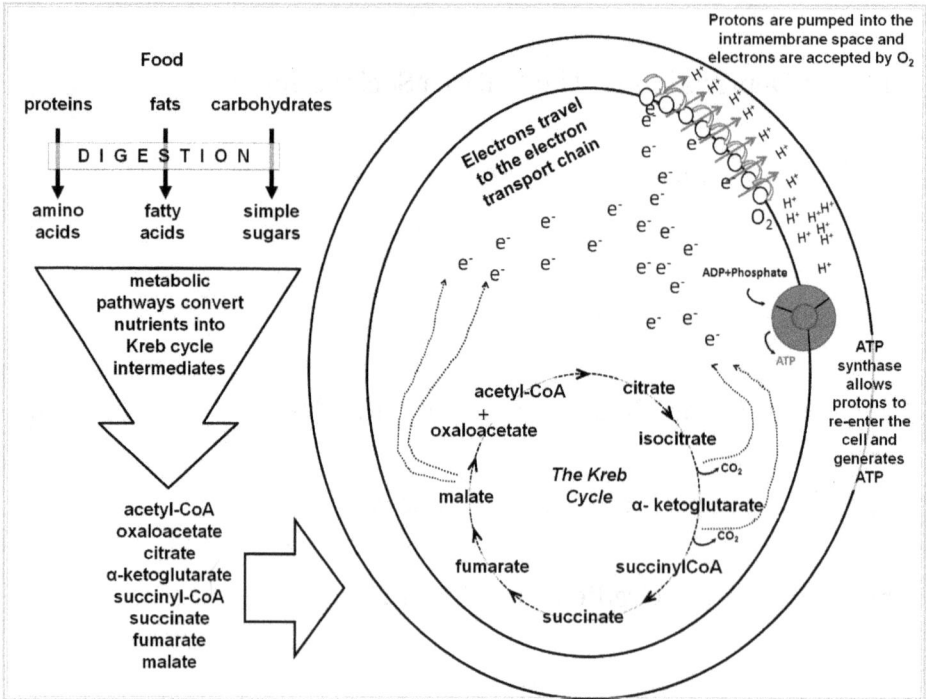

Cellular respiration is one of the most important concepts to understand in biology. It is an essential energy producing process that is key to maintaining life. You can usually tell how important something is based on what happens when it goes wrong. Interfering with cellular respiration at any one of its many steps can easily result in death. In addition to being one of the most important concepts in biology, it is also one of the most hated and feared among students. On a list of preferences for the average person, reading a book about cellular respiration usually ranks just below a colonoscopy or a trip to the dentist's office. Cellular respiration has everything that people generally avoid: diagrams with lots of arrows (we have one right in this chapter and it's hideous), complex words (oxaloacetate), complex words with greek letters (α-ketoglutarate), flashbacks to general chemistry (oxidations, reductions, and electrons, oh my), circular metabolic pathways (the Krebs cycle), proton gradients, acronyms (ATP), and confusing abbreviations with hyphens (acetyl-CoA). The many deterrents to studying cellular respiration prevent people from appreciating its beauty. Once you learn these concepts it really is a gateway and a cornerstone to many diverse subjects not the least of which is medicine.

I think that the trouble with the subject comes down to presentation. Cellular respiration is a detail intensive, multistep process that is often presented in the form of a schematic. It is nearly impossible to generate interest from a schematic alone. If you can't generate interest in a subject then it is never learned. If you force someone to learn it then the subject is forgotten at the earliest possible convenience which makes learning it in the first place essentially pointless. The many times I have taught this subject I was always confronted with the challenge of how to make it interesting. It is like being a chef and having an ingredient that people don't like and having the challenge of making it delicious. How do you make a schematic more interesting? The secret may be to demonstrate it from a functional standpoint. I have known a lot of people who were obsessed with automobiles. These people could tell you every last schematic detail about a car and how it worked and could probably draw most of it from memory. If you asked them how they got interested in automobiles they would say that it was how fast a sports car could go, how tight a motorcycle could corner, or how much a truck could haul. That interest led to a further exploration of the details of automobile workings. None of them would tell you that they were in a library one day when all of a sudden they came across a book on engine schematics which they studied intently for years. After reading that book and memorizing all the details of the schematic he/she all of sudden realized that automobiles can do many cool things. The diagram and detail laden subject of cellular respiration often falls into this trap. However, this time things will be different because we will focus on function over details; so buckle up because we are going to take the Krebs cycle out for a spin, walk down some dark metabolic pathways, reduce oxidations, and derail some electron transport chains. Sure, we have a diagram or two and we don't skimp on details but this is a biochemistry book unlike any other because we pack a body count. The subsequent chapters are packed with wild dogs, scheming socialites, ecological disasters, drunken party girls, and misguided cannibals. Each story focuses on one aspect of cellular respiration and the consequences of inhibiting it. The rest of this chapter will give a background of the framework of cellular respiration so that you can refer back to after reading the rest of the stories to better understand the process.

The first aspect of cellular respiration to address is its objective. Why does this process even exist? The goal of cellular respiration is to produce a portable form of energy that can be used to do work in biological systems. The end product of cellular respiration is a molecule known as Adenosine Triphosphate (ATP). The energy used to do work is hidden in its high energy phosphate bonds. You can think of ATP as a rechargeable battery. ATP is the battery in its fully charged state. To release the energy in ATP you have to break one of its phosphate bonds, which leaves you with ADP and a phosphate. ADP is a spent battery that you need to put on the charger. Cellular respiration is an example of a biological "charger" in which electrical energy is harvested from nutrients (sugars, fats, amino acids) and used to convert ADP and Phosphate back into ATP. In general, you need to do a lot of work to stay alive. You use a lot of ATP to do that work and when that work is done you are left with a lot of ADP and Phosphate. The consequences of not doing this work are typically quite severe, and by "quite severe" I mean death. If you don't do this work then you will die; so you always need more ATP and cellular respiration is the process used to replenish your stock. If you can't replenish your stock because cellular respiration has been disrupted you are left with piles of useless ADP and death is imminent. Harsh? Definitely, but that's the way it is. Laws are laws and the one to which I refer is the second law of thermodynamics, which states that the disorder in the universe is always increasing. Why do you die? To live you must maintain order, but the disorder in the universe is always increasing so everything in it is in a constant state of decay. To prevent that constant state of decay, you need to do constant work to restore order which in turn requires a constant supply of ATP.

When I think of decay and the necessity of work I always imagine sand castles at the beach. To have a sandcastle, you must build it from nothing. It takes work to construct a sand castle and the more ornate and intricate you make it, the more work you do and the more tired you become. However, once the sand castle is complete, its lifetime is limited as it can dry up, crumble and blow away or eventually get destroyed by the waves. You can try to maintain it by putting patches of sand in areas that get washed away or to replace dried and crumbled parts, but that takes work. Eventually the castle requires too much effort to maintain and falls apart when you get too tired or the decay happens more rapidly than you can fix. The sand castle can be thought of as a model for any biological system. All systems tend towards disorder and chaos, but to live you must create and maintain order which requires continuous work. To do work, however, requires energy. As such, it follows that the more energy you have, the more work you can do and the higher degrees of order you can maintain. You can bet that it takes a lot more ATP to maintain your body which consists of billions of cells than it does for a single celled organism like a bacterium to maintain itself. If you can understand all of this then you know why ATP is so important to produce, the purpose of cellular respiration, and the fatal consequences of its deficiency.

Now, how about the process of cellular respiration and how it generates ATP from ADP and Phosphate? Cellular respiration is a multistep process that occurs in cellular organelles, known as mitochondria. When you look at it from the start it seems very complicated, but if you can understand how a water wheel works then you can understand mitochondrial function. Water wheels are designed to harness the energy of flowing water. Water flows past the wheel and spins it. The energy of spinning is used to do other kinds of work, like milling grain or producing electricity (hydroelectric power). The stronger the force of the flowing water the more the wheel will spin and the more work will be done. ATP synthase is a protein that converts ADP and Phosphate into ATP; part of it functions exactly like a water wheel. The more the "water wheel" component of ATP synthase spins the more ATP is produced; however, the energy for the turning motion does not come from a river of water but one of hydrogen ions. As long as the hydrogen ions are flowing from the intramembranous space into the mitochondrial matrix, the wheel portion of ATP synthase will spin. The more water in a river, the stronger its flow so the more hydrogen ions in the intramembranous space the stronger its flow as well.

How do all those hydrogen ions enter the intramembranous space in the first place? They were pumped into that space by the motion of electrons through the electron transport chain. As electrons move through the transport chain, hydrogen ions are pumped from the mitochondrial matrix into the intramembranous space. The more electrons pass through the electron transport chain, the more hydrogen ions will be pumped into the intramembranous space, the more the wheel of ATP synthase will spin, and the more ATP will be produced. The electron transport chain is like a highway, electrons need to enter and move through it but they also have to leave. Electrons are carried to it, pass through it, and leave it when they are picked up by oxygen. This is why you need oxygen in your body. As long as oxygen is present, electrons will pass quickly through the chain and the energy of their motion will be used to pump hydrogen ions into the intramembranous space. If there is no oxygen, then electrons enter the chain and never leave which causes a traffic jam in this "electron highway." If there is no electron flow then hydrogen ions never get pumped into the intramembranous space, and there is no force to push the wheel of ATP synthase resulting in decreased levels of ATP and eventual death.

Where do the electrons come from? They are harvested from the intermediates of the Krebs cycle by a series of sequential oxidation reactions that form a cycle. Oxidation reactions are electron stealing reactions while reductions are electron donating reactions. The result of several oxidation reactions in this sequence is the generation of the carbon dioxide that you exhale. As long as intermediates are present in the Krebs cycle, then electrons can be harvested from them to feed the electron transport chain so that the hydrogen ion flow can be established to spin the wheel of ATP synthase to provide the energy for ATP generation.

Where do the intermediates of the Krebs cycle come from? Metabolic pathways are chemical assembly lines that convert nutrients(sugars, fats, and amino acids) into Krebs cycle intermediates. This is why you need to eat. When you consume food the carbohydrates, fats, and proteins are degraded into simple sugars, fatty acids, and amino acids respectively. Metabolic pathways convert these nutrients into any one of the Krebs cycle intermediates: acetyl-Coenzyme A(CoA), oxaloacetate, citrate, isocitrate, α-ketoglutarate, Succinyl-CoA, Succinate, Fumarate, and Malate. Metabolic pathways are often nutrient specific and there are a lot of them. Their details are far beyond the scope of this text, in this case the general goal is to focus on the purpose of a few of them.

Finally, to recap the entire process from start to finish: you eat food which your digestive system degrades into sugars, fats, and amino acids; these nutrients are converted into Krebs cycle intermediates by various metabolic pathways; electrons are stolen from the intermediates by the cyclic oxidations of the Krebs cycle which also results in the production of carbon dioxide; these electrons are then carried to the electron transport chain and passed down it where they are ultimately accepted by oxygen. The passage of electrons through the chain results in hydrogen ions being pumped in to the intramembranous space and their flow back to the mitochondrial matrix is the force used to spin ATP synthase and convert ADP and phosphate into ATP. Now that we have described the process let us see some cases where it is disrupted and witness some death and destruction with a side of α-ketoglutarate.

DRUNKOREXIA

Saturday night, and the club was packed. Brenda Devereux and her fellow interns at the investment bank, Claire Yee and Lily Palmer, were having their usual girls' night out. All three were wearing their tightest jeans and their tallest shoes and doing Bacardi 151 shots to celebrate the end of the week. "Brenda, you look so skinny," Claire gushed. "How do you do it? These drinks have a ton of calories, and you never gain an ounce."

Brenda smirked, tipping back another shot. "My secret."

"Let's go dance," Lily suggested. Brenda shook her head, earrings swaying. "I need another one of these before I can get out there and shake my fat behind." Lily made a face; she had less patience than Claire for Brenda's constant fussing about her weight. After another round of shots, they hit the dance floor.

"Excuse me, Porky," Brenda snapped suddenly as a short, curvy girl brushed past her. The girl glared at her, and Claire laughed. "Okay, you wanna know how I did it?" Brenda said loudly so the girl could hear. "I just don't eat before I go out. Like, anything. I haven't eaten for four days, so I can drink as many high-calorie drinks as I want, and I don't have to worry about bulging over my jeans like some people." This time—it was a couple of shots later—both her friends laughed. The girl they were mocking looked disgusted and went to dance where she couldn't see them. "That's right, cow, we didn't want to look at you anyway," Brenda giggled, and they all tipped back another round.

Brenda laid her shot glass down on the counter, and it rattled loudly. "Whoa," Lily said, "you're shaking. Are you okay? Maybe we better make this our last one."

"No, I'm good, I'm good," Brenda slurred. "It's still early. Just a little head rush is all. Whoa, where did you guys go?"

The other two women, drunk as they were, looked at her with concern. "We're right here, Brenda," Claire said. "We didn't go anywhere."

"Maybe she needs to throw up," Lily offered.

"Maybe she just needs another drink," Claire suggested. "Miss, could you get us another round?"

The bartender raised her eyebrows. "No," she said, "but I can get you a cab."

On the way out to the cab, Brenda could barely keep on her feet. The walk to the cab left her out of breath, and she kept complaining that her heart was pounding a mile a minute. They all got out at Claire's building and managed to wrangle Brenda upstairs with the help of the long-suffering doorman. "She just

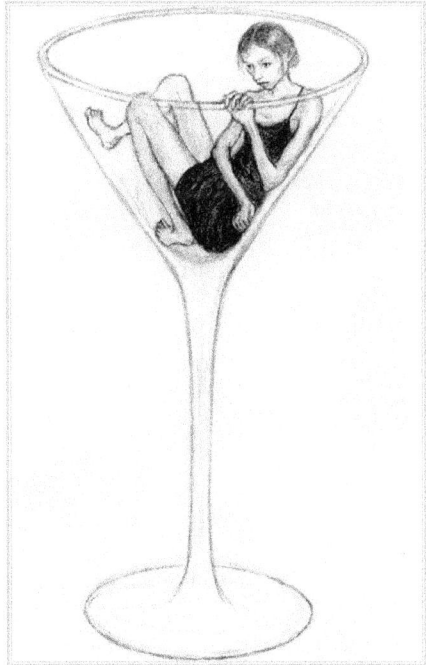

needs to sleep it off," Lily suggested.

"Well, you guys should stay here. Just stay here. Don't leave me alone with her, Lily, please? She's freaking me out."

That was at eleven at night. At two in the morning, Lily and Claire were awakened by the sound of something heavy falling. Claire fumbled for the light switch. Brenda had fallen off the couch and was lying, twitching, on the floor. As her friends watched, still half-drunk and confused, the twitching stopped. Lily knelt down next to her. "Brenda. Brenda." Brenda didn't move.

"Oh god," Claire whimpered, and Lily whispered, "I think we need to call an ambulance ... I can't feel a pulse ... I think she might be dead."

Scientific Connection

You, of course, are much too smart to do what killed Brenda, but more than one person has died this way. Sometimes referred to as "drunkorexia," the practice of avoiding food (fasting) so that you can consume alcoholic beverages at a later time is likely to give your night out a deadly ending. Because of its effects on glucose homeostasis, alcohol (ethanol) can be a lethal poison depending on how long it has been since the last time you ate. The amount of ethanol required to kill in this case would not be excessive and may be significantly less than is required for getting drunk. The smaller the individual, the less ethanol would be required to produce the lethal effects.

Glucose homeostasis is essential to the functioning of the brain. Glucose is the primary nutrient used by the brain for energy (synthesis of ATP). Sufficiently low blood glucose (hypoglycemia) will cause the brain to die by literally starving it of energy. Security is tight in the brain and glucose is one of the few molecules that can make its way in from the blood. In the brain, glucose is effectively the sole source of electrons harvested by the Krebs cycle and passed through the electron transport chain in oxidative phosphorylation. No glucose means no oxidative phosphorylation and certainly no ATP. The cells of the brain that conduct electrical signals are known as neurons and they work extremely hard. In order to do this work they need a constant supply of ATP, which means a constant supply of glucose as well. If their ATP production drops then they will stop working and die rapidly. Even short periods of hypoglycemia can lead to irreversible brain damage or death. The brain controls and coordinates the vital functions of the body like respiration and how fast your heart beats. If the brain dies then these vital functions will cease and the rest of the body will die as well.

You may have heard someone complain that they're irritable or have trouble focusing because of "low blood sugar," meaning they haven't eaten in a while. Under normal circumstances, your blood "sugar" levels never really change and are kept within strict limits. Insulin prevents your blood glucose from ever getting too high while the counter-regulatory hormones (adrenaline, cortisol, and glucagon) prevent it from getting too low. If you are fasting, the counter-regulatory hormones maintain your glucose levels by causing a constant release of glucose into your blood from the liver. Two processes are responsible for maintaining your blood glucose during a fast: breakdown of glycogen stores (glucose stored for a fast) and gluconeogenesis (synthesizing glucose from amino acids and other sugars). The body uses its glycogen stores, which can last between 10 and 36 hours, before turning to gluconeogenesis—the only option once the stored glycogen is gone. Ethanol interferes with gluconeogenesis, so consuming alcohol can lead a rapid drop in blood glucose if an individual's glycogen stores have been depleted.

An individual like Brenda who fasted for four days would have exhausted her glycogen stores long ago. By the time she started drinking she would be solely dependent on gluconeogenesis to maintain her blood glucose levels (and brain health). Consuming a beverage that was almost entirely ethanol would inhibit gluconeogenesis very quickly and cause a precipitous drop in blood sugar that could be lethal. The dizziness, anxiousness, jitteriness, sweating, rapid breathing and increased heart rate that Brenda exhibited were caused by the counter-regulatory hormone adrenaline, which was secreted into her blood in a vain attempt to get her liver to break down glycogen and release glucose into the blood. When Brenda's hypoglycemic episode continued, she lost consciousness and suffered seizures—signs that her brain was beginning to die and that even if doctors had been able to revive her, she might have suffered permanent brain damage. Eventually her starved brain could no longer control her vital functions and she was pronounced dead. This story is all too common, especially among younger drinkers. Usually individuals with alcoholic hypoglycemia are brought to the emergency room late at night and are very sleepy from the effects of alcohol intoxication. The next morning the sleepiness abruptly breaks as the affected individual goes into a seizure.

⇝ *Take Home Message* ⇜
Consuming alcohol after a prolonged fast can result in hypoglycemia which can destroy the brain and result in rapid death or lasting damage.

THE FALL OF A SOCIAL CLIMBER

FROM THE PROVIDENCE JOURNAL: July 9th, 2011
PROMINENT NEWPORT CITIZEN FOUND UNCONSCIOUS

NEWPORT, RI. Newport resident Florence Van de Camp was found unconscious this morning in her Bellevue Avenue home, "Crossways", by her husband,
Brock Van de Camp. A bottle of whiskey was found spilled on the floor near her
body. She was rushed by helicopter to Rhode Island Hospital and has not yet re-
covered consciousness.

Van de Camp, 64, inherited Crossways from her family, the Olens, who have
maintained a mansion in Newport since the early 1900s. Florence has continued
her family's practice of charitable giving, with an emphasis on entrepreneurship
and education; more recently, she has concerned herself with questions of healthy
body image among young women. She is a member of multiple organizations, including the Junior League, the South County Equestrian Society, and the Yacht
Club. She married Brock Van de Camp, now 38, in 2006.

Mr. Van de Camp stated, "Poor Florence has been so unhappy these last couple of years," he added, "She was drinking heavily again and refused to eat. Every
now and then I would find a liquor bottle hidden behind a book."

Mrs. Van de Camp remains unconscious in Rhode Island Hospital, where Mr.
Van de Camp visits her every day. The rest of the time, he can be found at the Fair
Winds Country Club, where he is working on his tennis game, "trying to keep busy
and take his mind off the tragedy."

FROM THE PROVIDENCE JOURNAL: July 15th, 2011
VAN DE CAMP MAY HAVE BEEN POISONED, EXPERTS SAY

NEWPORT, RI. Medical investigation suggests that the comatose state of
Florence Van de Camp, the well-known charitable giver and woman-about-town
found unconscious in her luxurious Newport home on Saturday, may have been
deliberately induced. Mrs. Van de Camp's condition deteriorated rapidly over the
last few days and she passed away from complications due to brain damage on July
11th.

In attempting to diagnose and stabilize Mrs. Van de Camp's condition, at-
tending physician Dr. Amos Charles noted a precipitously low blood glucose level
as well as unusually high levels of insulin in her bloodstream, but unusually low
levels of a molecule known as C-peptide.

"In the body, Insulin, the hormone responsible for lowering blood glucose,
and C-peptide are always found together," Dr. Charles said in an interview yes-
terday. "If one is high, the other is high, and so on. If insulin is high but C-peptide
is low, something is very wrong. At first, given her history of alcoholism and anorexia, we thought that it might be a case of alcohol induced hypoglycemia. Now I
think it seems more like murder."

While insulin produced in the body is always accompanied by C-peptide, Dr.
Charles explained, synthetic insulin is not. Mrs. Van de Camp is not a diabetic, but

these unusual blood test results made him curious enough to look for needle punctures where insulin might have been injected.

Mr. Van de Camp assured our reporters that to his knowledge, his wife had not been prescribed any intravenous drugs or other substances since her hospitalization for anorexia two years ago, but given her perilous emotional state she could have been using anything in an attempt to find peace.

FROM THE PROVIDENCE JOURNAL: August 3rd, 2011
NEW EVIDENCE IN VAN DE CAMP CASE

NEWPORT, RI. Investigators have revealed new, potentially incriminating evidence in the case of Florence Van de Camp of Newport. Mrs. Van de Camp's husband Brock has been detained on suspicion of poisoning his wife with insulin supplied by Christina Mazzucchelli.

Mr. Van de Camp, an avid tennis player, has frequently been seen in the company of Ms. Mazzucchelli—a former competitor in the U.S. Women's Open and now coaching tennis at the Fair Winds Country Club—since several months before his wife's fatal incident. Ms. Mazzucchelli has Type 1 Diabetes Mellitus and must inject synthetic insulin daily to keep her blood glucose levels low. The injections that keep her alive can kill someone whose body produces its own insulin and whose blood glucose does not need to be lowered.

Investigators have also questioned Ms. Mazzucchelli but have not yet determined whether she was a deliberate accessory to the crime.

FROM THE PROVIDENCE JOURNAL: January 18th, 2012
BROCK VAN DE CAMP CONVICTED

PROVIDENCE, RI. A jury has convicted Brock Van de Camp, 39, of Newport, of murdering his wife, Florence Van de Camp, by injecting her with insulin.

Under cross-examination, Van de Camp revealed that he and Mrs. Van de Camp, who inherited many millions from her family, signed a pre-nuptial agreement guaranteeing him 60% of her fortune upon her death unless he was unfaithful to her during her lifetime.

"But if she caught me playing around, I got a flat thousand," Van de Camp told the prosecuting attorney indignantly, adding that he had hoped his wife's history of alcoholism and anorexia would explain her death. The jury was out for seven minutes before asserting their confidence in Mr. Van de Camp's guilt.

Dr. Amos Charles testified to the presence of insulin and absence of C-peptide in Mrs. Van de Camp's blood, and explained for the court the mechanism by which insulin lowers glucose levels. His testimony is believed to have been essential in securing a conviction. Christina Mazzucchelli, from whom Mr. Van de Camp stole the insulin that induced his wife's coma, also took the stand during the trial but has been cleared of any deliberate participation in the crime. "Chris didn't know what I was doing," Mr. Van de Camp claimed. "I snuck her syringe out of her bag." Ms. Mazzucchelli was not available for comment. Sentencing will take place tomorrow.

Scientific Connection

Dr. Charles would have explained to the court that the brain is dependent on a constant supply of glucose from the blood for proper functioning and immediate survival. Glucose is the primary source of energy for the brain and its deprivation, even for a few minutes, can lead to tissue destruction and fatal losses of brain function. The brain is essential to survival because it controls vital functions like heart rate and the ability to breathe. If the brain dies then the rest of the body will soon follow. Survival therefore depends on maintaining a constant blood glucose level.

Blood glucose levels are maintained by the actions of two different sets of hormones. Insulin is responsible for lowering blood glucose after a carbohydrate-rich meal while the counter-regulatory hormones (glucagon, adrenaline, cortisol) are responsible for raising it during fasts. If too much insulin is present in the blood, then blood glucose can drop to lethally low levels.

Florence van de Camp's blood glucose dropped so far that part of her brain died. The regenerative potential of neurons (the cells responsible for the electrical signal conduction in the brain) is clinically nonexistent. Enough of Florence's brain was left functioning so that she could still breathe and her heart could beat fast enough to keep her alive, but the potential for conscious thought and movement was destroyed. Her husband injected her with more insulin during one of his hospital visits to destroy the rest of her brain and complete the murder he botched the first time.

Insulin is normally produced and secreted into the blood by the pancreas. However, the insulin that is made in the pancreas is a much larger molecule than gets secreted into the blood. Before insulin is secreted part of it—the C-peptide—is cut off and gets pumped into the blood along with, but separately from, the insulin. If Florence's own body had produced the insulin that caused her brain damage, her C-peptide levels would have been as high as her insulin levels.

People with Type I Diabetes Mellitus do not produce enough insulin to reduce their blood sugar to a safe level so they are required to take insulin by injection to maintain normal blood glucose levels. The type of insulin used to treat Type I Diabetes Mellitus is synthetic and does not contain C-peptide, so the fact that Florence's blood had a lot of insulin but very little C-peptide indicates that she had been injected with a high dose of artificial insulin. Since no one shoots insulin for fun, this led investigators to suspect that her coma was a result of an attempted murder. Brock relied on the town gossip about his wife's anorexia and alcoholism as well as a planted liquor bottle [see the "Drunkorexia" story] to suggest that her death was due to alcohol-induced hypoglycemia, leaving him fee to start a new life as a merry widower with Christina Mazzuchelli—on Florence's money.

Though this is a pretty twisted scenario, it is very similar to an actual case of hypoglycemia that put Sunny von Bulow, a wealthy socialite, into a coma for 28 years before she died. It is widely believed that her husband Klaus precipitated her coma by injecting her with insulin. The incident led to several highly publicized trails that were dramatized in the film *Reversal of Fortune*. You can read more about this saga of horror and hypoglycemia here: http://en.wikipedia.org/wiki/Sunny_von_B%C3%BClow.

~ *Take Home Message* ~

Insulin overdoses can precipitate hypoglycemia and cause serious brain damage and death. Insulin should be treated with respect and never misused.

THE CASE OF THE NOT SO STINKY DIAPER

Robert Jenkins, Jr., lay on his back on a soft white sheet. He moved his arms and legs tentatively, as if he wasn't sure if they worked, and stared at the faces leaning over him. "Does he know who I am, Mommy?" his older sister Mary asked.

Pamela smiled exhaustedly at her energetic seven-year-old. "He will real soon, especially if you do what you promised me before he was born. You remember how we practiced on your dolly? Show me how you can do it."

Mary carefully lifted Robbie's legs, undid the old diaper and wiped his behind with it. "Remember, don't try to throw it out right away," Pamela cautioned. "Put it to the side and get the new one on him first. This baby belongs to the whole family. It'll be up to all of us to take care of him."

"I know, Mommy," Mary said scornfully, already unfolding the new diaper and tucking it around Robbie's underside. "Wait'll you see," Mary assured her mother, closing the lid of the diaper pail. "You won't ever know he soiled his diaper. You won't even have to smell it. I am Chief Diaper Changer Mary Jenkins reporting for duty."

Pamela giggled and praised her daughter, mentally thanking Heaven that this was one task she wouldn't have to do. Her husband—dubbed "Big Rob" by his friends as soon as they'd picked the baby's name—had sworn to take on nighttime diapers and feedings. Unfortunately soon after Robbie was born he was forced to go out of state to work on a big contracting job. The work paid extremely well but forced him to live apart from the family for several days out of the week, and when he came back he was too tired to do anything. Little Robbie seemed to be keeping a mostly nocturnal schedule, requiring nighttime feedings but sleeping for most of the day. After the exhausting nights Pamela was extremely grateful to have Mary to help change Robbie's diapers in the afternoon while she got some much needed rest.

Seven-year-olds don't always keep their promises, but Mary was as good as her word, playing or watching TV near her brother and whisking him away to the changing blanket whenever it was necessary (and sometimes when it wasn't). "You won't ever know" was a bit of an overstatement, but Pamela rarely changed any diapers at all. On Saturday, Pamela's parents came over for dinner as they always did. An hour after Robbie was fed Mary gave him his evening changing. Her mother and grandma worked on dinner and Big Rob, who just returned home from a

month of work, watched TV with his father-in-law, who didn't even know how to turn on a stove. "Your mother tells me you're a great big sister, Mary," her grandma said. "I hope Robbie isn't too much trouble for you." Pamela rolled her eyes. Mary said quickly, "He's such a good baby, Grandma. He's so sweet, even his diapers smell sweet."

"Isn't she an angel," Pamela's mother cooed. Pamela smiled too and said, "That's pretty clever, Mary."

"What does clever mean again?"

"Um—well—in this case it means I think you made a good joke."

"It's not a joke. They really do. Not when he does number two, but his pee smells really sweet, like pancakes, and sometimes his skin does too. I think he might be magical."

"Like pancakes? Mary, is this something you saw on TV?"

"No!" Mary was getting perilously close to a tantrum. Why didn't grownups ever believe anything kids said? "Wait here, I'll go get his last diaper" A few minutes later she called out, "Okay, Mommy, smell," and held Robbie up for Pamela to take. Pamela hooked a finger in the back of the diaper, sniffed, and said, "I just smell poop, Mary. Nothing sweet."

"Let me." Mary took him back and sniffed too. "It doesn't smell like it now. But sometimes it does, especially when he hasn't been fed in a while." Noticing Pamela's expression, she shouted, "You think I'm making it up but I'm not!"

Pamela turned to her mother. "Ma, have you ever heard of anything like this?"

"No, dear," her mother said vaguely. "We'd better call the men in—the chicken's just about ready."

Later that night, Pamela asked her husband, "Rob, have you noticed a weird smell coming from Robbie's diapers?"

Big Rob laughed. "That wouldn't be weird, honey."

"I mean weird for a diaper. Like ... sweet."

"You think I smell them? Usually I'm doing my best not to smell them."

Pamela was strangely unsettled by Mary's observations and decided to do the afternoon changings herself from now on. Mary was an honest kid and not a particularly imaginative one; she always wanted to know how things really worked. Pamela's apprehension increased as the afternoon drew near. When it came time for Robbie's changing, Pamela nervously unsnapped Robbie's onesie and sniffed his diaper. A sweet smell met her nose. Had she somehow bought scented diapers by mistake? But this didn't smell like Fresh Linen or Spring Rain or any other marketing chemical; it smelled like maple syrup. "Pancakes," Pamela said out loud. Robbie gurgled. She thought, "Maybe it's nothing. Maybe lots of people's pee smells like syrup—how would I know? It just doesn't seem natural, though." She sat with Robbie a while, stroking his tiny hands. Then she unearthed her phone from the couch cushions and called the pediatrician.

"It's a good thing you came in. This sounds like it could be Maple Syrup Urine Disease," said Dr. Lillian Bit, putting a Band-Aid on Robbie's thigh where she had drawn blood.

Normal
Metabolic
Disease

"Mary's the one who noticed it," Pamela felt compelled to say, but added defensively, "I change his diapers too, but I didn't notice this."

"Hmm. Were you mostly changing him after feedings? If he has a mild version of the condition, and it seems like he does since his neurological exam is fine and the smell is not always present, the symptoms might not show up if he has been recently fed. The disease manifests when certain branched chain amino acids are burned to make energy, this can occur when an excess of them are ingested or during fasting when muscle proteins are broken down into amino acids for fuel. Since these branched chain amino acids are present in almost every protein the generalized break down of proteins into amino acids to be used for fuel could precipitate the signs of this disease. It is likely since his metabolism is through the roof that all amino acids in his diet would be used for protein synthesis and thus would not be used for energy production. As he slept all day he would be fasting and so the smell would be present in his evening diapers but absent from his morning diapers after he had been fed several times during the night."

"So you're saying this is an actual disease."

"Oh yes, a silly yet descriptive name for a serious disease of amino acid metabolism that can cause permanent neurological damage and death. However, in more severe cases the smell is present more frequently as it is a sign of toxicity. Pamela, I'm 99% sure this blood I've drawn will confirm what I've just said. I'll give you some handouts to take home, about how to set up a diet and a feeding schedule that will keep Robbie from going into this state, and we should schedule a checkup between now and his next appointment—say two weeks from now. The doctor smiled. "Good job, Mary. Maybe you have a future as a physician specializing in Metabolism and Genetics."

Scientific Connection

Metabolic disease can be a complicated topic, but if you are capable of understanding the workings of an assembly line then you are more than capable of rationalizing the basis of metabolic disease. In a traditional assembly line, workers start with raw materials and each worker in the line does a specific job. As the product passes from worker to worker it is modified slightly until a complex product like an automobile is created. Assembly lines are as important to biology as they are to industry: proteins are the factory workers of biology and often work together exactly like this to convert one chemical, or several chemicals, into an end product. Any chemical that has been processed in this way is considered to be "metabolized". These chemical assembly lines are more properly called "metabolic pathways", and proteins that participate in them typically get their names from the job they do. These names are usually highly descriptive but unfortunately instead of "riveter", "roll turner" or "glass installer", they are a little more unwieldy, like "pyruvate carboxylase". However, just because someone has a job does not mean that they will be good at it.

If an assembly line worker is incompetent then partially assembled products will start to accumulate and back up, resulting in chaos.

A perfect example of this can be seen in this classic clip (http://www.youtube.com/watch?v=4wp3m1vg06Q) from I Love Lucy where Lucy and her friend Ethel work in a chocolate factory to disastrous ends. If a protein in a metabolic pathway is bad at its job or (even worse) not present at all, incompletely processed chemical intermediates will back up and accumulate until they reach toxic levels. Some metabolic diseases like Maple Syrup Urine Disease can be considered poisonings but instead of the poison coming from the outside world, it's made inside the person's body due to defects in a metabolic pathway. A person with a metabolic disorder can consume something non-toxic and because it cannot be completely processed its intermediates will accumulate and poison the individual. Another example of a metabolic disease is Hereditary Fructose Intolerance in which a simple glass of fruit juice can be lethal. Treatment usually consists of complete avoidance or minimal exposure to whatever it is that can't be metabolized. In a factory you can fire a bad employee and hire a new employee that will hopefully do a better job. In biology you aren't so lucky, because your body makes its own "worker" proteins based on the information contained in DNA. The segment of DNA that tells you how to make a specific protein is called a gene. In cases of metabolic disease, a gene of the affected individual has been altered or "mutated" so that the information on how to make a key protein is bad. This leads to the production of a protein that is horrible at the job it was originally designed to do. The sequence of genetic mutation, incompetent protein, and resulting disaster is the basis of metabolic disease.

In order for any nutrient (sugars, fats, amino acids) to participate in cellular respiration and be used to make vast amounts of ATP, it must first be converted into a Krebs cycle intermediate. In the case of poor little Robbie, Branched-chain α-ketoacid dehydrogenase complex—a key protein involved in the pathway that converts branched chain amino acids (valine, isoleucine, leucine) into Krebs cycle intermediates—is defective. This means that trouble is going to start whenever excess dietary protein is used to make ATP or when bodily protein is broken down for ATP generation during fasting states or illness. When any of these situations occur, incompletely processed intermediates that possess a sweet odor similar to maple syrup accumulate in the blood; when they reach high enough levels they will be excreted in the sweat and urine of the affected individual, thereby producing the characteristic odor. The maple syrup smell is often the signal that someone has this disorder and is being poisoned (many disorders involving amino acid metabolism have distinctive odors associated with them, most of which are really bad: "boiled cabbage", "mice", "tomcat urine"). Maple Syrup Urine Disease is not one specific disorder but rather a collection of disorders that vary in severity based on how badly Branched-chain α-ketoacid dehydrogenase complex is doing its job. In the most extreme and more classic case, Branched-chain α-ketoacid dehydrogenase complex can't do anything at all. In this most severe form the smell is present within the first week of life and signs of nervous system toxicity like poor feeding, rigid muscular paralysis, seizures, coma, and death follow within a week or two. Robbie has one of the milder forms of the disease in which the smell is occasionally present after fasting, because while his α-ketoacid dehydrogenase complex can't work as well as most people's it does still retain some activity.

The treatment is to prevent amino acids from being used in cellular respiration and limit the amount of branched chain amino acids in the diet. It is good for an individual with this disorder to eat fairly regularly and avoid fasting. As long as this individual does not eat anything containing excessive amounts of protein production of toxic intermediates will be minimized, so bread is good and prime rib is not. Robert will have to avoid both excessive amounts of protein in his diet and prolonged fasting for the rest of his life to avoid toxicity. Thanks to the vigilance of his big sister and the suspicions of his mother, Robert's case was caught early enough to prevent neurological damage. Fortunately, in modern medical practice newborn screening is performed within the first week of life to identify metabolic disorders so that they can be caught and managed before they can do serious harm. Physicians specializing in Endocrinology, Metabolism and Genetics manage diseases like this one so if you find this case interesting it could be a possible career path for you.

↬ Take Home Message ↫

Nutrients have to be converted to any of the intermediates in the Krebs cycle in order to be used in cellular respiration. Metabolic pathways are responsible for this conversion and defects in them lead to diseases.

The Krebs Cycle

WILEY COYOTES

Death goes hand in hand with life on a livestock farm. Before he turned five, Ian Henderson had seen his dad kill the male lambs for the table. But staring out over his station—the grazing ground of South Australia, the farm that was his now—he dreaded what he knew he'd find: the bodies of the sheep that were his livelihood, their throats and bellies torn out, their meat and fleeces useless.

Packs of wild dingo dogs had arrived in the early spring and taken out the lambs first—this year's meat and next year's wool. Then they started in on the ewes whose fleeces he was planning to sell. Henderson and three of his hired hands guarded the night pasture with rifles in hand; but you can't keep sheep penned up in the daytime, and you can't be everywhere at once. He'd left out poisoned meat and even poisoned an already-killed sheep; the dingoes ignored it. He'd set spring-traps powerful enough to snap a human leg; the dingoes strolled around them. His own dogs were good at herding sheep but no match for the vicious packs. Every night one or two fewer sheep came in; every day he found another corpse.

What was to keep the wild dogs from coming back till every one of his sheep—each worth a thousand or more—was dead? He began going into town more often for a drink, and then for two drinks, and then for five. Grief and rage boiled up in him. He poured more whiskey on top of it. The man at the other end of the bar said, "You seem thirsty, mate," in a deep mild voice.

Henderson had always believed that a man doesn't talk about his troubles, but what did he have to lose? The stranger sipped his beer, silent till Henderson stopped talking. Then he reached into his pocket and pulled out a small card. DINGO DESTROYER, it said, and a phone number. "Call him," the stranger said, and took another sip of beer.

The Dingo Destroyer pulled up in a battered truck three days later at sundown: a short, scruffy, balding man with a dufflebag over his shoulder. Henderson's heart sank. The man waved away his explanations and followed Henderson to the night pasture, where he opened his bag and began fastening collars around the thick, wooly necks of the sheep. Henderson opened his mouth to speak. "Fluoroacetate," the Dingo Destroyer said. "Poison."

"I tried poison."

The Dingo Destroyer ignored him. When all the sheep had collars—they were long, wide plastic packets, made to fasten—he said, "You been keeping a guard at night?"

"What do you think?"

"Don't do it tonight. Leave them out."

"You're crazy. You're—" But the little man

13

was already walking back to his truck. Henderson watched him drive off. He went back to the house. He told the hired hands to go on home; they looked at him like he was the crazy one. He went to bed early and lay there all night.

At dawn, all the sheep but five were dead, their throats torn out.

The stupid, worthless packets around what had been their necks were empty. Henderson thought he might pass out; he thought he might scream. He would find that—curses flew from his mouth—and do to him what he'd done to his sheep, corpse after corpse—

He stopped. The dead, stiff body at his feet wasn't a sheep; it was a dingo. So was the one in the corner. So was the one he had stepped right over, not even recognizing it, on the way into the pasture. Henderson counted twenty-two of the dead wild dogs. Looking at the five sheep left alive, huddled in the corner of the pasture, Henderson's rage turned into violent joy: his enemies were dead! Breathing hard, he gave the nearest tan body a furious kick. Then his eyes went to the five sheep that remained to him, and both rage and joy subsided into relief. He could start again.

Scientific Connection

Cellular respiration and farming have a lot in common. The goal of farming is to harvest something (corn, wheat, wool, etc.) so it can be sold for a profit. A large harvest will result in a gigantic profit and a thriving farm, while a poor harvest will result in a scant profit and a bankrupt farm. Believe it or not, the mitochondria in the cells of your body harvest electrons, so that they can be sold for energetic profit in the form of ATP.

You consume meat and vegetables that are purchased from farms. Your digestive system reduces these foods into simple nutrient molecules (simple sugars, fatty acids, amino acids) and absorbs them into your blood. The blood in turn carries these nutrients to the hardworking cells of your body. The mitochondria in these cells harvest electrons from the nutrients through a series of sequential oxidation reactions known as the Krebs cycle. The purpose of the Krebs cycle is to steal as many electrons from each nutrient as possible until all that remains is a completely oxidized product (carbon dioxide), which you breathe out. The more electrons a nutrient has, the richer the ATP harvest. Fats are loaded with electrons and more ATP can be made from them than an equal mass of sugar.

In order to be "harvested", nutrient molecules must first be chemically converted into one of the intermediates of the Krebs cycle (Acetyl-

CoA, Oxaloacetate, Citrate, Isocitrate, α-ketoglutarate, Succinyl-CoA, Succinate, Fumarate, or Malate). The electrons taken from these intermediates are then whisked away by electron carriers (NADH, FADH2) and passed down the electron transport chain until they are finally accepted by oxygen. The passage of electrons through the electron transport chain generates the proton gradient necessary to power ATP synthase, the enzyme responsible for generating ATP from ADP and phosphate (ADP + phosphate→ATP). In short, the more electrons you can harvest the more ATP you can make. If your cells make a continuous supply of ATP they will continue to live; however, if their rate of ATP production crashes they will die and so will you.

Fluoroacetate is a gigantic monkey wrench that gets thrown into the Krebs cycle and shuts it down hard. Inhibiting the Krebs cycle prevents electrons from being harvested. As a result, the other events involved in cellular respiration cannot occur and ATP production comes to a halt resulting in rapid cellular death. Fluoroacetate is metabolized to fluoroacetyl-CoA, which bears a striking resemblance to acetyl-CoA, a normal Krebs cycle intermediate. Fluoroacetyl-CoA is converted into fluorocitrate, which acts as a powerful inhibitor of, aconitase, the enzyme responsible for the conversion of citrate to isocitrate. The inhibition of aconitase stops the progression of the Krebs cycle and the rate of ATP production decreases to fatal levels. Fluoroacetate is a phenomenally effective killing agent and is widely considered to be one of the most toxic substances in existence. It is primary used as a pesticide, especially against wild dogs. Wild dogs like to bite the throats of sheep; as a result sheep are often given collars filled with fluoroacetate that cover their necks. Any wild dog foolish enough to bite one of these sheep will receive a lethal dose of fluoroacetate that will halt ATP production in nearly every cell of its body, resulting in certain death.

⇀ Take Home Message ↽
Inhibition of the Krebs Cycle can lead to fatal decreases in ATP production.

CYANIDE AND THE CASSAVA

It is the end of the 16th century and a group of Spanish explorers are on an expedition into the jungles of South America to find wealth and glory. The men entered the jungle in hopes of finding gold and treasure, but as the hazards of the environment cut them down one by one, dreams of material wealth are abandoned and the only goal is survival.......

The pale, sweating man stumbled and planted both hands in a pile of decaying leaves. The smell of the jungle was all around him. He thought he heard his name, but it must have been a monkey or a bird: none of his companions was alive to say it. Caimans and piranhas had gotten Vasquez on the way across the river. Javier had fallen to the poisonous dart-guns of three men who had not believed his story about being a god from across the sea. A jaguar had leaped on San Germano when he got up to relieve himself in the middle of the night. Now only he, Morales, was left alive to finish the expedition. But as starvation weakened his body and addled his mind survival had gone from fleeting hope to impossible dream. His fingers scrabbled the dirt and felt something like bark. Roots—they might be edible! He dug frantically, pulled the thick brown root out of the ground, and groaned in disappointment: it was the one that the people in the last village had told the foreigners never to eat. "It makes you sick and you lose control of yourself," they had explained through the interpreter, "and then you die." Morales learned this first hand as Miguel, another of his companions, had consumed it out of hunger. At first Miguel was fine but soon after growing weak and confused was left behind to die. A quick death due to poisoning seemed much better than being

mauled by animals or murdered by natives. Morales, smiling at the thought of being free of this nightmare and joining his friends, wiped the root on what was left of his tunic and tried a tiny nibble: bitter as jealousy and hard as an old boot. He spat the root out in disgust. Maybe stewing it would help. He could at least die with one last good meal. He had a tinderbox but nothing to cook in—their packs were at the bottom of the river with Vasquez's bones. He took off his iron helmet, dipped water from a nearby puddle, and built a fire under that. When the root was soft, he took it out and looked at it. If it killed him, his sufferings would be over. If he lived, he would make this miserable land pay. He took a bite—it was bland, but in his starving mouth it tasted better than his mother's home-baked bread. Hunger overcame him and he gobbled the rest of it down. His belly full for the moment, no longer caring if he lived or died, Morales went to sleep at the foot of the tree in a state of peace he had not known since entering the jungle. His first thought when he awoke was glee: I'm still alive! Day after day, Morales continued to make this stew until he was once again at full strength. In time he left the jungle realizing that life was more valuable than any amount of gold and anxious to share with the world the food that had saved his life.

Scientific Connection

The stew that Morales made is known today as tapioca, and this story of disaster and attempted suicide is one of the legends surrounding the discovery of this popular treat. Tapioca is a starch that is extracted from the cassava root. Starches are carbohydrate rich plant extracts that are typically used as thickening agents in cooking, similar to the corn starch that you can find in your local grocery store. However, unlike corn, the cassava can be loaded with cyanide. In many plants, including the cassava, cyanide is bound to chains of glucose forming what are known as cyanogenic glycosides. As long as the cyanide stays bound to those glucose chains you are safe, but if cyanide gets released from those sugars it will poison you. As a general rule, living things don't like to get eaten and the cassava is no different. When the plant cells of the cassava are crushed—by teeth, for instance— the destroyed cells release proteins that start to cut cyanide away from the sugar molecules, precipitating its toxicity.

Cyanide is a potent inhibitor of cellular respiration and works to inhibit ATP production by blocking the electron transport chain in mitochondria. Electrons are harvested from nutrient molecules (like sugars, fats and amino acids) by the Krebs cycle. These electrons are passed down the electron transport chain where they are finally accepted by oxygen. As electrons are passed through the chain, hydrogen ions are pumped into the space between the inner and the outer mitochondrial membranes. The generation of this hydrogen ion gradient is really the most important part of cellular respiration because the more hydrogen ions you can fit between the inner and outer mitochondrial membrane, the more ATP you can make. ATP synthase is like a little water wheel: as the hydrogen ions pass by it they push the wheel and make it spin. The spinning of the wheel is what generates the energy to make ATP from ADP and inorganic phosphate. Cyanide blocks the electron transport chain, preventing the flux of electrons and the formation of the hydrogen ion gradient, and leading to potentially fatal decreases in ATP production. It is similar to hydroelectric power plants, in which water wheels are pushed by gigantic waterfalls. The force of the waterfall spins the water wheels and the mechanical energy of those spinning wheels can be converted into enough electrical energy to power a town. However, if the powerful waterfall is reduced to a trickle the wheels will stop spinning and then no electricity will be produced.

ATP is stored biological energy that is used to do essential biological work. The harder the tissue must work, the more ATP is needed to keep it functioning. The tissues of the nervous system contain very hardworking cells called neurons that require a significant amount of ATP to function. While you cannot physically see decreases in ATP production, you can observe the effects of a tissue that is not working. Because cyanide inhibits ATP production, tissues like the nervous system that depend on high levels of ATP generation will suffer first. Acute cyanide intoxication, as in Miguel's case, causes confusion, agitation, and disorientation because it impairs the nervous system and prevents normal cognition (thinking). If the nervous system is significantly impaired, then death will follow because it controls the rate of breathing and the beating of the heart. As with the snake venoms in previous stories, when heartbeat and respiration stop, death follows quickly. Morales, on the other hand, seemed to be unaffected. What did he do that Miguel did not?

The secret lies in the making of the stew, Morales crushed the root up, dried it out, and boiled it in a pot. The drying and boiling process destroyed the protein that liberates the cyanide from the glucose chains and facilitated the elimination any of the volatile (easily vaporized) cyanide that had been liberated during the crushing process. By making the stew, Morales disabled the cyanide defenses of the cassava and made it safe to eat. The proper handling and preparation of cassava is a worldwide health issue, because it is a major crop in the Southern Hemisphere, especially Africa, which produces 30 million tons of cassava root/year. In rural regions of Africa, cassava represents a major portion and occasionally exclusive component of the diet especially in times of war, famine, and drought. Cassava sold in American grocery stores in has low cyanide content and is not nearly as hazardous to eat raw. In places where high-cyanide cassava is grown, improper processing and preparation can lead to outbreaks of serious illness. The chronic cyanide exposure that results from consumption of improperly processed cassava primarily affects the nervous system producing two well known diseases: Konzo and Tropical Ataxic Neuropathy.

PATIENT SKETCH OF KONZO

A young boy needs a cane to walk. Even with the cane he struggles considerably and makes almost a tripod with it as he hobbles along. Both legs are straight and move without a bend in the knees. The legs seem to be very weak.

Konzo means "tired legs" and it is heavily associated with a diet rich in high-cyanide-content cassava in the setting of low dietary protein intake. It results in an irreversible spastic (stiff) paralysis of both legs. It seems to strike children around the age of 2-3 or women of childbearing age. Your body, while not immune to cyanide (CN-), can detoxify it by converting it to thiocyanate (SCN-) but you need sulfur in your diet to do that, and dietary sulfur comes from foods that are high in protein. Cassava is a carbohydrate-rich food that is very low in protein and thus its exclusive consumption is a setup for this sort of scenario.

PATIENT SKETCH OF TROPICAL ATAXIC NEUROPATHY

A 70-yearold man with poor vision and hearing has a difficult time feeling things touch his skin. He has a hard time feeling pain and has burned himself a few times without even knowing it. He can walk if his eyes are open, but if he closes them and tries to walk he will trip even if there is nothing to trip over. With his eyes closed he is extremely unsteady and is at risk of falling over, but when he opens them he is considerably more stable on his feet.

This is tropical ataxic neuropathy, which is believed to be caused by many years of eating poorly processed cassava in the background of low dietary protein intake. Unlike Konzo, which results from higher amounts of consumption over a short period of time, tropical ataxic neuropathy is believed to result from lower levels of exposure for a much longer period of time. The nervous system is still affected, but in konzo the effect is paralytic and tropical ataxic neuropathy is more or less a destruction of the senses. Associated signs of tropical ataxic neuropathy: sensory neuropathy (the inability to feel physical sensations like being touched or burned), bilateral optic atrophy (serious decreases in the vision of both eyes), bilateral deafness (inability to hear in either ear due to damage to nerves), and sensory ataxia (inability to walk with eyes closed because all other senses that help you coordinate, like, touch are gone due to nerve damage. The ability to see where you are going corrects this).

↪*Take Home Message*↩
Cyanide inhibits cellular respiration and can cause fatal decreases in ATP production. Cyanogenic glycosides are found in many plants besides cassava, such as apple seeds and peach pits.

THE CURSE OF LAKE NYOS

It started because Nurse Christine Nkwain hadn't heard from her sister in nine weeks. "This is not usual," she said firmly to Karl Jacobs, the World Health Organization doctor at her clinic. "You always say you are here to help."

"I'm happy to give you the day if you want to go yourself," Karl offered. There were plenty of needy people right there in Bamenda; no need to go chasing after more. Nurse Nkwain was shaking her head, and there was an unfamiliar expression on her face. After a year in Cameroon he still had trouble reading people's body language, but he thought she looked afraid. "I don't go there," she said. "We only write letters. It's a bad place."

"The village?" He cast his mind back over his briefing; he couldn't recall any warnings about problems, political or medical, in the village of Cha. "The lake. Lake Nyos is a bad place." When she saw his dark eyebrows go up she added, "I'm a Christian, Dr. Jacobs, you know that. But there are evil spirits in that lake, or why would people die just from being there? When my sister married a man from Cha, I mourned for her as if she were dead. I'm always afraid until I get an answer to my letters. Nine weeks is too long; I'm sure the spirits have her."

"People die just from being there? That's not possible, Nurse. There must be some sickness endemic to the area that hasn't been diagnosed. You have medical training; how can you embrace those ignorant superstitions?" He realized he'd gone too far as her face froze into a mask of contempt. "But," he added, "if there's sickness there it could spread. I'll investigate it tomorrow morning; I'm meeting with the schoolteachers today about immunizations. Please get those materials ready for me and be ready to translate into French or Pidgin if necessary." Jacobs himself spoke English well, French badly and Cameroonian Pidgin not at all.

The next day, after getting a later start than he had hoped for, Karl loaded up his bike basket with food and water, water purification tablets and quinine pills, and basic diagnostic and first-aid equipment; it would take him most of the day to reach Cha. Though August is wintertime south of the equator, it was hot, and it seemed to him that the outline of the Oku Volcanic Field barely changed as he pedaled along the mountain road. It was empty except for him—no trucks, no bikes, no one on foot in either direction—and he began to zone out as he rode, so that when his front wheel hit the dead frog he almost fell.

He stopped to make sure the tire hadn't blown out but saw, instead of a nail or a sharp rock, the tiny corpse. He frowned. Surely a frog could hop out of the way of a bike? His confusion, and his sense of unease, increased as he continued toward the valley where Cha was: every few feet, it seemed, he saw a dead rat, a dead lizard, another

dead frog. They weren't roadkill; there were no wounds or marks of crushing; strangest of all, nothing had eaten them. This was an all-you-can-eat buffet for vultures; where were they? And then he saw them, or at least two of them—lying dead by the side of the road. "Evil spirits," he said out loud, "what superstitious garbage!" His voice sounded strange to him, and the muscles in his legs didn't seem to want to pedal. He forced himself to bike onward; his shadow grew longer as the sun sank lower; he could see the roofs of Cha in the valley below, while the hill that held Lake Nyos loomed off to the left. Here on the slopes were the villagers' crop fields and grazing grounds—and the bodies of antelopes, monkeys and foxes. He forced himself forward around a bend in the road and stopped cold, swearing and cursing.

An entire herd of cows—fifty or more—lay on the grazing ground where they had fallen. They were beginning to rot, and so was the boy who had been watching them, but no scavenger had touched them. Karl leaned over to the side of the road and vomited. His bike had fallen to the ground; his mind was empty of everything but panic. "There are no evil spirits," he said out loud; his voice sounded like an insect buzzing in the bushes, but there were no insects buzzing. After what might have been a minute or an hour, he picked up the bike and walked it beside him down into the village itself. There was no sound of grindstones scraping or goats bleating; no sound of children fighting or playing; no sound of women and old men humming or chatting while they worked.

They were all dead. The people of Cha, like their cattle, lay where they had fallen—over the grindstones where they had been making fufu, over the babies they had been nursing, near the doorway they had been mending. One woman lay across the fire where she had been cooking. Even the flies were dead. There was no sign of what had killed them—no marks, other than those of decay, on their bodies. The smells of so much death sickened him and he vomited again.

When he looked up, he looked again at the woman lying dead across her cooking fire. Her body and clothes should have been burned, but they looked like all the others. To keep from panicking, he began to look around at the other cooking fires and coal stoves. Unattended, they should have burned themselves out, and the indoor ones would probably have set fire to the roofs. Instead, they seemed to have stopped burning long before their fuel was used up. He used his radio to send out a call for help. What could have caused such devastation?

Scientific Connection

This is a dramatization of "the Lake Nyos disaster", one of the worst environmental catastrophes in history. Four different villages were affected and over 1,800 people and thousands of animals were killed. Whatever the lethal agent was, it acted so quickly that there was no opportunity for escape. What could kill so many, so quickly, and in such a wide area?

The legends about Lake Nyos were correct about its danger but not its lethal mechanism. The absence of burning lamps and candles in any of the four affected villages was an important clue to unraveling the mystery. Fire is visible evidence of a combustion reaction. Combustion reactions can only occur if oxygen is present to accept electrons from whatever is being burned. No oxygen means no fire. This evidence led investigators to believe that all of the oxygen in the valley had been rapidly eliminated or displaced for some period of time. It is believed that a gas cloud of tremendous size and density displaced air from the valley, but what kind of gas could do this and where would it come from?

In sufficient quantities, carbon dioxide gas is more deadly than any evil spirit. Carbon dioxide gas is extremely dense; a physical property that allows it to be poured like a liquid. Concentrated carbon dioxide gas could displace the ambient gases of air (20% oxygen, 80% nitrogen) from the valley just like poured water would displace the air that occupies the inside of a cup. An estimated 1.6 million tons of carbon dioxide gas was effectively poured out of Lake Nyos and into the village below it. The volume of carbon dioxide gas was so gigantic that it pushed the air out of the valley, resulting in the rapid death of nearly every living thing in it.

The theory behind the source of the carbon dioxide gas had to do with several factors: the great depth and stagnancy of Lake Nyos; and the activity of a nearby volcano. Lake Nyos is one of the world's deepest lakes and it is also one of the most stagnant. Continuous, mild volcanic activity beneath the lake produced significant amounts of carbon dioxide gas that accumulated in the water until it was completely saturated.

Carbon dioxide gas seeps into every lake from the ground but the motion of the water brings it to the surface so it never accumulates to dangerous levels. The water of Lake Nyos is very stagnant so massive amounts of carbon dioxide gas were able to accumulate within it as the water lay undisturbed. It is believed that every now and then a big bubble of carbon dioxide gas would come to the surface and asphyxiate any nearby organisms, thereby leading to the legend of the soul-stealing demons. In the case of the Lake Nyos disaster, a massive cloud of carbon dioxide gas was released rather than a gigantic bubble.

The precipitating event for the release of the gas cloud is unknown, though it is thought to be related to an acute increase in volcanic activity. This could have caused a giant shock wave to pass through the water, forcing the normally stagnant water to move rapidly. This massive shake-up of lake water could have release carbon dioxide gas in the enormous amount required to overwhelm the valley.

Oxygen deprivation is extremely dangerous. Oxygen is the most physiologically important gas in air due to its role in cellular respiration. Electrons are harvested from nutrient molecules (sugars, fats, amino acids) by the oxidation reactions of the Krebs cycle. These electrons are then passed through the electron transport chain in order to generate the energy necessary to establish a hydrogen ion (proton) gradient. The energy of the proton gradient provides ATP synthase with the necessary power to make ATP (ADP + Phosphate→ATP) in the same fashion that water wheel harnesses the power of a waterfall to make electricity.

Much like a highway, the electron transport chain is all about flow. Electrons have to enter and electrons have to leave. As long as things are flowing everything is fine. Oxygen accepts electrons at the end of the chain and allows them to leave, just like an exit ramp allows you to leave a highway. Without oxygen, electrons continue to enter the transport chain but do not leave leading to saturation or an "electron jam". The rate of ATP production grinds to a rapid halt without oxygen to accept electrons from the electron transport chain. Insufficient production of ATP prevents cells from doing work that is essential to maintaining order. Once order is lost the cells will die rapidly. Imagine a highway where cars continued to get on and none ever left. In time the whole highway would fill up with cars and all traffic would stop. Oxygen is essential to removing electrons from the transport chain so new electrons can enter and lead to a high ATP production rate.

⇀Take Home Message↽

Oxygen is an essential component of cellular respiration. A constant supply of oxygen is essential to produce enough ATP to keep you alive.

DNP DIET DISASTER

Daniel Johnson is a 24-year-old accountant who is significantly overweight. He is looking to vacation at the Jersey Shore this summer, and he wants to be a big hit on the club scene and attract a lot of women. Daniel believes that unless he gets a set of abs like his idols on the "Jersey Shore" his chances of getting any attention from the ladies are slim to none. He figures that he has to lose at least 30 pounds in 2 months to achieve this physical goal. While surfing the internet one night Daniel sees an ad for diet pills called "ADIMAX". The drug is being marketed as "too extreme" for the FDA and is only to be used by people who are "serious about losing weight fast". Daniel ends up ordering several boxes of ADIMAX. It takes a few weeks for the pills to get imported from Russia but as soon as they arrive at his apartment he tears the packaging open like a kid on Christmas morning.

He notices that there is no recommended dosage information or anything about active ingredients. The only thing on the cover of the bottle besides the word ADIMAX are pictures of people with ripped abdominal muscles. Daniel is not sure how many pills to take at one time, but he figures that he is serious about losing weight so he should take at least ten because that seems like a nice round number. He maintains this regimen for a few days and begins to lose weight. One day his co-worker, Craig, notices that Daniel's shirt is so soaked with sweat that it is transparent. Daniel seems to be breathing much faster than normal and he is clutching his chest like a person might do if their heart were racing. Daniel tells Craig that he feels like the blood is boiling inside of his body. When Craig touches Daniel's skin it is hot to the touch. Craig knows that this is not natural and takes Daniel to the emergency room where it is determined that his body temperature is 104°F (normal 98.6°F).

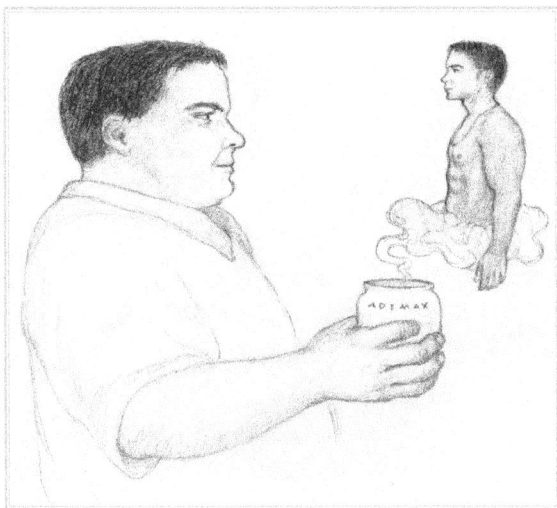

Scientific Connection

This is just one example of the many risks associated with idolizing cast members of the Jersey Shore. Going up to strange women, pulling up your shirt, and showing off your abs will lead to more restraining orders than phone numbers. Unregulated dietary supplements exist in a spectrum between useless and deadly. The best way to lose weight is to adopt a philosophy of healthy living that includes modifications to your diet and increases in exercise. The internet is a breeding ground for dangerous "quick fix" products that use "modern science" to yield impossible results. These products often do not list things like "ingredients", though the buyer is often assured that the components are "all natural" (as we've seen, cyanide is also natural—and deadly). In this case, the fictional product ADIMAX contains a drug known as dinitrophenol, which was used in diet pills in the 1930s. Dinitrophenol acts as an uncoupler of mitochondrial oxidative phosphorylation and is banned in the United States for use as either a dieting aid or a pesticide. The scientific rationale behind its use as a diet aid is logical, but the benefits are greatly outweighed by its potential for danger.

Uncoupling agents target the mitochondria, the organelle responsible for ATP production by oxidative phosphorylation. In the mitochondria's innermost region, the matrix, electrons are harvested from nutrient molecules by the sequential oxidation reactions of the Krebs cycle. Electron carriers (NADH, FADH2) taxi electrons to the electron transport chain, which they flow through until they are accepted by oxygen. The passage of electrons through the transport chain provides the energy for hydrogen ions to be pumped into the space between the inner and outer mitochondrial membranes, resulting in the formation of a hydrogen ion gradient. The hydrogen ions really want to make their way back into the mitochondrial matrix but they only have one passageway that lets them do this: the pore of the protein ATP synthase. ATP synthase is like a little revolving door: as hydrogen ions make their way through it part of the protein spins. The force of this turning motion provides the energy for the crucial reaction that makes ATP (ADP + Phosphate →ATP). The formation of the hydrogen ion gradient is essential to the spinning of ATP synthase and the production of ATP. The more ATP synthase spins, the more ATP gets made.

An uncoupling agent like dinitrophenol interferes with the formation of the proton gradient. Electrons are still passed down the electron transport chain and accepted by oxygen; unfortunately, the hydrogen ions leak out of the intramembranous space as quickly as they are pumped into it. In order to increase the amount of protons pumped into the intramembranous space, more electrons are passed through the transport chain, which ironically leads to increased oxygen consumption but decreased ATP production.

How does all of this result in weight loss? The electrons used to generate the hydrogen ion gradient are harvested from nutrients like fat inside of the mitochondria. The uncoupling action of dinitrophenol requires more nutrients to be burned to make the same amount of ATP, which indeed would lead to weight loss. Unfortunately the increased flux of electrons through the electron transport chain leads to the deadly side effect of dinitrophenol: elevated body temperature.

Your average body temperature is 98.6°F, but where does this heat come from? Why are living people warm and dead people cold? The answer lies in the mitochondrial production of ATP. The process of passing electrons down the electron transport chain generates most of the heat that warms your body. When your body dies you stop producing ATP, which means that you stop producing heat as well. With no heat production the body cools to room temperature. In the case of poisoning with dinitrophenol, there is increased flux of electrons through the electron transport chain. More electrons being passed through the chain means more heat is being produced and the body temperature will rise so high that it can be life threatening. In order to meet the increased demands for oxygen and nutrients the heart beats faster to deliver more blood to the tissues. Breathing rate is also increased to acquire more oxygen from the outside world.

↪ *Take Home Message* ↩
The hydrogen ion gradient is the force that powers ATP synthase and the generation of ATP. It is created by the passage of electrons down the electron transport chain which also generates heat. Uncoupling agents interfere with the formation of the hydrogen ion gradient and can cause both decreases in ATP production and extremely high body temperatures.

THE CURIOUS CASE OF THE CORN CANNIBAL

THE SCENE: A courtroom in Providence, RI. Doug West, a 46-year-old molecular biologist, is on trial for multiple murders in the first degree.

DR. JULIA YANG'S TESTIMONY

I am the head of the molecular biology research team at Hyperion Technologies in Cranston, Rhode Island. Dr. Doug West was one of five applicants who made the first cut for a spot on this team. His CV revealed brilliant graduate work on genetically modified organisms, and his first job was in that field with Gene Frontier Laboratories, an affiliate of the Monsanto Corporation. The CV also revealed a troubling employment gap of five years, from 1998 to 2003. Naturally, I asked him about this time period. He began to fidget, raised his voice and said that he had been "developing his dark gifts." When I asked him to be more specific, he became extremely agitated, stood up and started pacing around the room and shouting. He told me that he was "an intellectually gifted cannibalistic spirit of the forest." He referred to himself several times as a "Wendigo" and said that a dark being known as the "the Nameless One" had revealed to him "the forbidden secrets of knowledge." Although he wasn't threatening me, I judged that he might become dangerous and pressed the buzzer to call security repeatedly until they arrived. He said that during those five years he had been "following the dictates of the Nameless One to gain great power" and that this made him the best candidate for the job because "he could achieve things that ordinary scientists could only dream of." At this point, the building security arrived and restrained him while we waited for the police. They took him away and I haven't seen him until today.

DETECTIVE MARK BRONSON'S TESTIMONY

I'm a Detective Second Class with the Rhode Island State Police. The Yawgoo Valley Police called us in because in additional interviews after his arrest, West boasted that in order to get his "extraordinary powers" he had "absorbed some of the best scientific minds in recent history." Then he said some names: Radhika Gupta, Laura Van Damme and Michael Wozzeck. Those are three scientists who worked at the Gene Frontier Laboratories until they disappeared. Gupta vanished in 1999, Van Damme in 2001 and Wozzeck in 2003. This made us take another look at the cold case file: West was a suspect in the disappearances because in a staff meeting in 1998, he lunged at Dr. Gupta and bit her arm deeply enough to draw blood. Like you might expect, West was dismissed and when Gupta disappeared we thought it might be an act of retaliation for the dismissal, but we found no evidence and had to put the case on ice. His mentioning those names made us think there might've been a connection after all and we got a warrant to search West's home and its grounds.

On July 9th, 2004, myself, Detective Sergeant Romney and Officer Alvarez drove to West's home, which is near Taunton in the area known as the Bridgewater

Triangle. West lives on two acres backed by the Hockamock Swamp and most of those two acres are planted in corn. I don't know anything about farming, but it looked like regular corn to me. We entered the house and saw no evidence of anything unusual until we reached the kitchen. There were two brand-new refrigerators and every inch of them was crammed with ears of corn and sticks of butter. This seemed pretty weird—I mean it seemed unusual enough to justify further investigation of the cornfield. Those same two officers and I came back the next day along with a backhoe operator. We excavated the cornfield and buried underneath it we found the skeletal remains of three human bodies. Forensic examination and comparison with dental records proves that these were the bodies of Gupta, Van Damme and Wozzeck. Because West kept insisting that he was a "Wendigo" and that we would be sorry when he achieved his full powers, the District Attorney advised that we get a psychological evaluation and called in Dr. Freeman.

DR. LOUIS FREEMAN'S TESTIMONY

I'm the consulting psychologist brought in by the State of Rhode Island to evaluate Dr. West's mental state. When we met, Dr. West introduced himself as a "Wendigo". He informed me that he had researched my career and didn't think I was prestigious enough to be worth "absorbing." West told me that in 1998, he received a visit from the "Nameless One" who revealed to him his destiny as a superior being and instructed him to "absorb the powers" of others by eating their flesh, which would make him the greatest molecular biologist in the world. This,

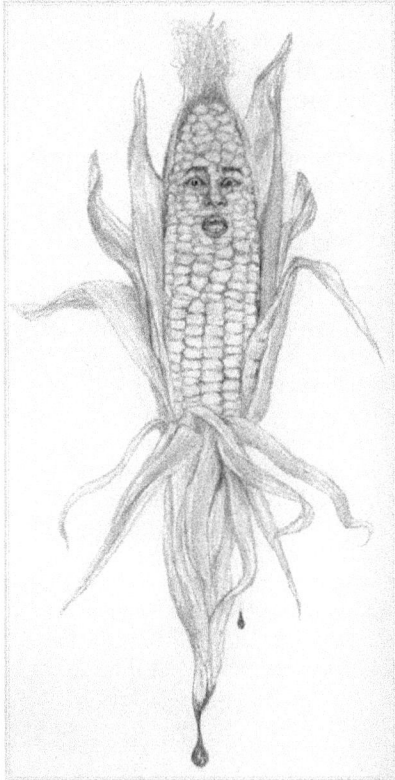

he said, was why he bit Dr. Gupta's arm, but he found he didn't like the taste of human flesh. He had worked with corn in the past and concluded that the best way to consume his colleagues would be to allow their dead bodies to fertilize the corn. In decaying, he said, the complex molecules in their bodies would break down and enter the corn stalks as simpler molecules that he could then digest, making their "powers" his own. West appeared calm unless I made any attempt to challenge his version of reality. I concluded that he was suffering from persistent delusions and hallucinations. Let me explain those terms to the jury: a delusion is a fixed false belief that can't be shaken by the presentation of contrary evidence. A hallucination is a sensation that does not come from any verifiable exterior source—the most common example is "hearing voices." These symptoms strongly suggested to me that Dr. West is the victim of a possibly temporary but probably permanent psychosis—a loss of contact with reality—and thus was not able to distinguish right from wrong at the time he committed these crimes.

DR. DOUG WEST'S TESTIMONY

I don't know why Dr. Freeman keeps talking about "psychosis." I'm not crazy; I had a vision. The Nameless One told me that I was destined to be the greatest molecular biologist in history, but I couldn't do it alone. I needed the powers that my colleagues had. Sure, it was fine to kill them. They didn't have any destiny. I did just what the detective said. I'm not ashamed of it; I killed them, I buried their bodies in the cornfield, and I ate the corn that grew from their bodies. But you know, it hasn't worked so far. I can't feel myself getting smarter, and I think I know why. I should've done my math more carefully. See, the main thing that plants want from organic material in the soil is nitrogen, and only 3% of the human body is nitrogen. The rest is 65% oxygen, 18% carbon, 10% hydrogen—those are the big ones. If I assume that the intellect of the scientists was distributed equally among all elements and that the corn absorbed all of their nitrogen, I would've had to eat the entire corn stalk, not just the ears, to gain just 3% of their intellect. Maybe there's no shortcut, maybe I would really have had to eat all three of them, but I just couldn't stand the taste of Dr. Gupta's flesh and I couldn't face eating 500 pounds of that. I mean, that would just be nuts. 500 pounds of corn is a totally different story. I just love the stuff! You know, now that I think about it, if I had studied botany instead of molecular biology I wouldn't be in this mess at all.

Scientific Connection

You might think that this story is implausible, but it pales in comparison to many actual cases involving psychotic individuals. Many people who have committed a crime try to fake a psychosis on the legal grounds that a psychotic person is not responsible for his or her actions, but the aberrant behavior exhibited by psychotic individuals is so extreme that this strategy is rarely successful. Although many people who commit violent crimes are sane enough to know right from wrong, Doug West's hallucinations and delusions suggest that he has indeed lost contact with reality; in addition, his plan would not have worked. He should have stopped listening to evil spirits and picked up a biology textbook.

Dr. West is correct that matter is made of elements that are passed from organism to organism throughout the biosphere. In living systems the most common elements are carbon, hydrogen, oxygen, and nitrogen, with limited amounts of phosphorous and sulfur. This elemental recycling is easily approximated by "The Circle of Life" analogy made in the Lion King. When a gazelle eats grass, it absorbs the elements the grass is made of and incorporates some of those elements into its own molecules. The same thing happens when a lion eats the gazelle. After the lion dies, decay will break its complex molecules into simpler ones that are absorbed by and incorporated into plants, scavengers, and bacteria. When another gazelle eats those plants, the cycle continues.

This constant recycling of matter throughout the ecosystem has given rise to "matter cycles" like the carbon, nitrogen, and water cycles. The elements in your body have been recycled over and over through again for trillions of years. Some of your carbon could have at one time been in the bodies of dinosaurs, saints, philosophers, and kings.

Dr. West is also correct that not all of the scientists' molecules would have entered the corn and then his own body. Unlike herbivores and carnivores, which consume organic material (meat or plants) to obtain carbon for the synthesis of complex molecules, plants obtain the carbon they need from the carbon dioxide gas in the atmosphere, which is the product of cellular respiration in animal and bacterial cells. This carbon dioxide gas is captured or "fixed" by plant cells to be used in photosynthesis, which in turn produces the oxygen essential to cellular respiration. Simply put, this means that these two principles oppose one another in their actions and complement each other's effects. Because plants get their carbon from the air, they would not absorb the carbon in the bodies of the murdered scientists. Plants do absorb nitrogen from the soil (many artificial fertilizers contain high percentages of nitrogen), but nitrogen composes only 3% of the human body—making this system of indirect cannibalism extremely inefficient. In addition, intellect and ability are not chemicals and aren't passed on through the carbon, oxygen or nitrogen cycles; if they were, you would only be as smart as the cows or carrots you consume. So if you were thinking of eating the kid next to you in order to do better on your next test, forget about it.

⇝ *Take Home Message* ⇜
Matter can be neither created nor destroyed and is constantly recycled throughout the biosphere.

FRAMING ANALOGY - BATTERIES NOT INCLUDED

Christmas morning, 1987, 5:40 a.m.

Six-year-old Billy Walton squirmed in his bed and looked at the clock. His parents had told him the night before, firmly, that Santa doesn't come until 6 in the morning, and if he went downstairs to get his presents before then, the elves would take them back to the North Pole. But they just said he couldn't go downstairs. They didn't say he had to be asleep—and who could possibly sleep when the Terror Tank 8000 was waiting downstairs?

It was the toy of his dreams: a remote control tank that could climb walls, jump over holes, and fire both lasers and missiles at the same time. The first TV commercials for it came on the day after Halloween while Billy was eating his left-over candy and watching G.I. Joe. The day after that, Billy wrote the first of 54 let-ters to Santa—one for every day between Halloween and Christmas Eve—request-ing a Terror Tank 8000. Some letters he placed in the mail; others he gave to any white-bearded, chubby old man he encountered, in case this was Santa in disguise, checking (as he always threatens to do) who's naughty and who's nice. He even stopped picking on his little brother, knowing that he and the Terror Tank 8000 could make up for it later.

Six o'clock! Billy leapt out of bed and catapulted downstairs. One box was bigger than any of the others under the tree; as his parents and little brother stum-bled sleepily after him, Billy tore the wrapping off the glorious Terror Tank 8000, pried the box open and the toy from its wrappings, and pressed the button on the remote control.

Nothing happened.

While Billy burst into tears, his mother checked the label on the box to see what might have gone wrong. "Uh-oh," she said. "Batteries not in-cluded. We need 20 D batteries—18 for the truck and two for the remote control."

"It's not a truck, it's a tank!" Billy wailed.

"Hey, settle down," his dad said. "Settle down. Look at this, buddy, I came prepared." He produced two 10-packs of D batteries. "Help me get these in there."

"Mike, you didn't buy those at Hardware-R-Us, did you? I got an ex-tension cord there the other day and it blew out my hairdryer." Together, Billy and his father installed the bat-teries and tried again to begin the Terror Tank 8000's first mission of mayhem. But the Terror Tank 8000 didn't climb the wall; instead, it bumped against it. It couldn't leap over the sink; it just fell in. It would go forward, but only slowly. The

laser lights were dim and the missiles, instead of soaring across the room like in the commercials, just dropped. "Dad, why isn't it working?"

"I'm sorry to say that I think your mom is right—sorry, babe, that came out wrong. What I mean is that I bought these batteries for cheap. They look fine, but it seems they don't work so well. Just to be sure, let's check them in something else." He rummaged in the kitchen drawer and pulled out a flashlight, which worked fine with its present batteries; with the new batteries, though, its light was dim. "Well, buddy, there's your answer. The tank just can't get enough energy out of these cheapo batteries. The good news is that when stores are open again, we can get some regular batteries and the Terror Tank 8000 should be able to ride again."

Scientific Connection

Energy is the capacity to do work; the more energy you have the more work you can do. Of course, if you don't have any energy then you can't do any work. If there is an energy failure in a system, the observable sign is that elements of the system stop working or don't work as well as they should. In this case the work that Billy wanted to see was the fast movement, glowing lasers, and firing missiles of the tank. However, because the batteries didn't have enough energy stored in them, the tank could not do all of the cool things that it was supposed to. It moved, but slowly instead of fast; the laser lights were dull instead of bright; and the missiles didn't really work at all. In the same way, parts of a biological system working poorly or not at all can be a sign of an energy-source problem. The mitochondria are the battery of the cell: it is responsible for producing energy so that cellular work can be done. People with mitochondrial disorders have a set of bad batteries in pretty much every cell of their body. Usually tissues that require lower amounts of work to function tolerate this better, but tissues that require a lot of work to function (heart muscle, skeletal muscles, and nerves) suffer severely.

Mitochondrial disorders fit into a spectrum depending on how many diseased mitochondria the sufferer has per cell. This means that two people with the same disorder can have totally different disease processes based on the proportion of diseased mitochondria to healthy ones. Some people are barely affected while others die in childhood. Just as with the Terror Tank 8000, the proportion of good batteries to bad ones determines the overall functioning: more good batteries than bad means better functioning, and the opposite is also true.

One other fact makes mitochondria stand out: mitochondria are responsible for replicating themselves, so they have their own DNA. All of the mitochondria in your body originate in the oocyte, which is the genetic information you get from your mother's side. The proportion of good to bad mitochondria in that oocyte determines how severe the disease will be in your body. Typically mitochondrial diseases cause problems with multiple systems of the body; as noted above, the most striking abnormalities are associated with the muscles and nerves because those tissues require a lot of energy to function. Here are some examples.

MITOCHONDRIAL DISEASE 1

You are an ophthalmologist. A 20-year-old woman comes to your office and wants to know why she is losing her vision. In the past few months her vision has progressively decreased in both eyes and now she is nearly blind. Questions reveal that she's not the only one in her family whose vision has suffered: she has two brothers who both went blind much earlier in life. Her mother had progressive loss of vision, but never went completely blind.

Scientific Connection

This is a mitochondrial disease known as Leber's hereditary optic neuropathy (LHON). The target tissue is the nervous system, as evidenced by the degeneration of the optic nerve and the development of blindness. This disorder most frequently presents as a loss of vision in both eyes leading to blindness at a young age. This vignette also captures the spectrum of the disorder, with two brothers affected much earlier than their sister and a mother who has a slightly milder case. This means that the mother inherited fewer of the oocytes with diseased mitochondria from her mother; the early blindness of the two boys means that out of everyone in the family, they inherited the greatest percentage of diseased mitochondria per cell. The variability is due to the different numbers of diseased mitochondria in the oocytes that gave rise to each child. All the children are affected, as is the pattern of inheritance with mitochondrial disorders.

MITOCHONDRIAL DISEASE 2

A normally healthy nine-year-old girl is having an after-school snack at the kitchen table when she realizes that she can't move her right arm and leg. She wants to call for help, but she can't speak. She gets her family's attention by knocking a glass off the table with her left arm; they rush her to the hospital, where doctors determine that she has had a stroke. When tested, her serum lactate is 5.6 mM (normal is <1.5 mM) and questions about her family's medical history reveal that she had a little brother who died suddenly before his first birthday.

Scientific Connection

This is a case of MELAS syndrome: Mitochondrial myopathy, Encephalopathy, Lactic Acidosis, and Stroke-like episodes. Like LHON above, it is caused by mitochondria that do not work as well as they should. The mitochondria cannot make enough energy to satisfy the work needs of the nervous system.

This lack of energy caused the little girl's stroke. A stroke occurs when the cells of the brain don't have enough energy to keep working. The result is a loss of function, like the inability to move the limbs on half of her body and the loss of speech. She has a high blood level of lactate because the only way to make energy outside of mitochondrial mechanisms (oxidative phosphorylation) is glycolysis. In the absence of working mitochondria, lactate is the result of glycolysis, which is why the doctors tested for it in the case of this stroke. High levels of lactate can acidify the blood and disrupt the electrical workings of the nervous system and the heart. Her little brother also had this disorder, but he had more diseased mitochondria per cell and died of a lethal stroke at a very young age.

MITOCHONDRIAL DISEASE 3

You are a pediatrician. Your newest patient is a young boy whose guardian tells you that he has been periodically falling, twitching and losing consciousness. The boy is in the tenth percentile for height and shows signs of mental impairment. He has progressively lost his hearing and is now deaf, and your tests reveal that his vision has been deteriorating as well. The boy has difficulty walking and coordinating motions; his reflexes are perfect but his muscles are universally weak. You take a sample of his muscle tissue and examine it: under the microscope the mitochondria look clumped together and the muscle cells are ragged, as though they have been ripped apart.

Scientific Connection

This is a case of MERRF: Myoclonic Epilepsy with Ragged Red Fibers. It is a mitochondrial disease that typically presents with seizures, ataxia (inability to coordinate muscle movement), deafness, vision loss, muscle weakness, stunted growth and poor mental development. The tissues with high-energy demand, like the nervous system and the muscles, are heavily affected. The loss of hearing, loss of vision, seizures, poor mental development, and loss of coordination are all evidence of energy failure in the nervous system. The weakness of the muscles and their torn appearance under the microscope is evidence of energy failure in muscle tissue. The muscle cells don't have enough energy to do their job and survive so they choose to do their job and die, resulting in their ragged, ripped appearance.

MITOCHONDRIAL DISEASE 4

A woman brings her young son in to your office because of frequent and re-curring infections. The boy is short for his age. His mother says that he is winded every time he goes upstairs and that he tires easily in general. Your questions reveal that she had a brother and an uncle who had similar symptoms. A blood test shows extremely low numbers of white blood cells, which explain his frequent infections.

Scientific Connection

Barth syndrome is another mitochondrial disease. It's character-ized by increased risk of infection, rapid fatigue and constant feeling of being tired, delayed growth, learning disabilities, and poorly working heart. Death usually is due to heart failure or infection. Once again, this disease shows a pattern dominated by energy failure in muscle tissue. As a result there is generalized muscle tissue weakness, evident in both the skeletal muscles and in the heart. Unlike the previously discussed mitochondrial disorders, the inheritance pattern associated with this is X-linked recessive, which is why the mother's uncle and brother are af-fected as opposed to her and all her children. While mitochondria are responsible for replicating themselves and making most of their own proteins, some of their proteins are made from information in the nucle-us, which accounts for this pattern of inheritance.

MITOCHONDRIAL DISEASE 5

You are a first-year medical student doing a physical on a woman in a hospi-tal. When you look into her eyes you notice an unusual color in both of her retinas. You ask her to follow your finger with her eyes and she can't move her eyes from side to side. When you ask her to stand up and walk to the other side of the room, she does so with a wide unsteady gait like she is drunk. Her muscle strength is poor and she has a history of heart problems.

Scientific Connection

This woman has Kearns-Sayre Syndrome. It is characterized by abnormal retinal pigment accumulation, opthalmoplegia (inability to move eyes from side to side), myopathy (generalized muscle weakness), ataxia (poor coordination), heart problems, and diabetes. This disease has all the hallmarks of mitochondrial dysfunction including reduced function of the nervous system, skeletal muscles and heart.

Take Home Message: Mitochondrial diseases cause energy failures in multiple tissues. The tissues with high energy demands suffer the most dysfunction. The skeletal muscle, cardiac muscle, and nervous tissue are the most frequently affected. The degree of severity depends on the proportion of diseased to healthy mitochondria per cell.

Authors and Illustrators

Do you like action? Adventure? Intrigue? How about Cellular Respiration? The Providence Alliance of Clinical Educators brings you a text book unlike any other you will ever see. We take the exciting process of Cellular Respiration out of the pages of so many dry textbooks where it has been quietly rotting away and put it into furious action. The pages of this book are packed to the brim with forlorn Conquistadors, Scheming Socialites, Villages of the Dead, Dark Rituals, Vicious Dingoes, Drunken Party Girls and a baby whose diapers refuse to stink. You want the Krebs cyle? We got it. You want oxaloacetate and acetyl-Coa? They are here too. The chapters of this book are designed to present Cellular Respiration in a functional manner that will give you an in depth understanding of the process presented in a style that is truly visceral and geared towards self directed learning. The concepts of Cellular Respiration are presented in the context of stories so that you can see the principles in action, what we create are fables for the scientific age. What is the danger of an insulin overdose? What happens if you disrupt the krebs cycle? Why do we need oxygen? What is a mitochondria? The answers to all of these burning questions and more are just a page turn away. If you are only going to buy one textbook about Cellular Respiration this year then buy this one twice!